D0031864

Fire in Oakland, California

Titles in the *American Disasters* series:

Fire in Oakland, California

Billion-Dollar Blaze

Carmen Bredeson

AMERICAN DISASTERS

Enslow Publishers, Inc.

40 Industrial Road   PO Box 38
Box 398   Aldershot
Berkeley Heights, NJ 07922   Hants GU12 6BP
USA   UK

http://www.enslow.com

Library of Congress Cataloging-in-Publication Data

Bredeson, Carmen.
    Fire in Oakland, California : Billion-dollar blaze / Carmen Bredeson.
        p. cm. — (American disasters)
    Includes bibliographical references (p. 44) and index.
    Summary: Describes the devastating fire that swept through
Oakland, California, in 1991 and the experiences of firefighters,
police officers, and ordinary citizens during this disaster.
    ISBN 0-7660-1220-4
    1. Oakland (Calif.)—History—20th century—Juvenile literature.
2. Fires—California—Oakland—History—20th century—Juvenile
literature. 3. Oakland (Calif.)—Biography—Juvenile literature.
[1. Fires—California—Oakland.] I. Title. II. Series.
F869.02B74   1999
979.4'66053—dc21   98-31284
                                                          CIP
                                                          AC

Printed in the United States of America

10 9 8 7 6 5 4 3 2

To Our Readers: All Internet addresses in this book were active and appropriate
when we went to press. Any comments or suggestions can be sent by e-mail to
Comments@enslow.com or to the address on the back cover.

Illustration Credits: AP/Wide World Photos, pp. 1, 8, 9, 13, 15, 17, 19, 20, 21,
22, 24, 27, 30, 31, 32, 33, 35, 36, 37, 38, 39, 40; © Corel Corporation, p. 6;
Reproduced by Enslow Publishers, Inc., p. 12.

Cover Illustration: AP/Wide World Photos.

Contents

*T*he Oakland Bay Bridge, connecting the city of Oakland to downtown San Francisco, was completed in 1936 and spans 2,310 feet.

CHAPTER 1

Sunday, October 20, 1991

"It was a wall of fire. The wall of fire was 100 feet high. The wind was pushing the flames at fifty miles per hour. The fire had jumped the freeway. People were fleeing from their cars. Cars exploded. Houses exploded," June Jordan recalled.[1]

That morning, before she ever saw the fire, Jordan sensed that the air seemed heavy. She called to her dog, Amigo, but the dog would not move. Jordan said, "I noticed his hind legs quivering, and I thought, 'Earthquake!'"[2] Running around to the front of the house, Jordan saw a wall of fire bearing down on her Oakland, California, neighborhood.

On another street in Oakland, Kristine Barrett-Davis and Mark Davis noticed that their cat was acting funny. Disney was turning around and around in circles and meowing loudly. Within minutes a black cloud of smoke settled on their home, and the couple knew what was bothering Disney. Quickly they grabbed a few belongings, picked up the cat, and headed for the car. On their way,

Disney jumped free and ran. Kristine and Mark frantically tried to catch the cat but finally had to leave without her. Soon afterward, fire swept in behind them and burned their house to the ground.

Flames were already leaping one hundred feet into the air by the time the fire approached Betty Ann Bruno's house. Embers, blown by the wind, had landed on a neighbor's wood-shingle roof and set it ablaze. Bruno

Many homes were destroyed during the Oakland fires when hot embers came in contact with roofs made of dry wood shingles. A roof of this type can be seen on the house at the right.

Kate Taufa and her daughter, Emily, watch flames engulf homes on an Oakland hillside where Kate's parents' home is located.

grabbed a garden hose and started spraying water on the roof of her own house. As she sprayed, the sky got darker and darker. Soon a firefighter ran by, yelling at people to evacuate. Bruno took a last look at her home and left. Later she said, "The fire was too fast. My house was burning down within fifteen minutes of when it hit the neighborhood."[3]

In the Oakland hills, across the bay from San Francisco, people were in a state of panic. Many jumped into their cars and tried to escape down the narrow, hilly roads. Some were stopped by a wall of fire and had to abandon their cars and run. When fire reached those cars, the gas tanks in some exploded and the tires burst. Many people who tried to run through the flames were burned, some fatally. These fatalities were among the twenty-six people who lost their lives in the one-day fire.

Animals were also fleeing in panic. Herds of deer crashed through yards, along with terrified raccoons, rabbits, and skunks. Family pets ran for their lives, too, as flames licked at their fur. Many pet owners were not at home to rescue their dogs and cats, so the animals had to escape on their own. Some were trapped inside burning houses and could not get out.

The sound of sirens filled the smoky air in the Oakland hills as dozens of fire trucks converged on the scene. Firefighters jumped from their trucks and began connecting hoses to water hydrants. Their job was complicated because the fire was spreading so fast. How had this fire gotten out of control so quickly?

CHAPTER 2

The Beginning

On Saturday, October 19, 1991, a brushfire broke out in the heavily wooded hills behind Oakland and neighboring Berkeley, California. Firefighters put out the six-acre blaze in about two hours. They left some hoses at the site in case any hot spots developed later. The next morning, several firefighters returned to the scene of the fire and found a few places where embers still burned. Just as these small fires were being put out, the wind picked up and started to blow very hard.

What happened next started one of the worst fires in California's history. Captain Donald Parker of the Oakland fire department said, "Eyewitness accounts testify that a sole ember blew into a tree just outside the burn area, and the tree exploded into flame, and the resulting fire was quickly out of control—raging around and over firefighters who were indeed fighting for their lives."[1]

The fire spread so fast and was so intense that firefighters were forced to retreat. They immediately

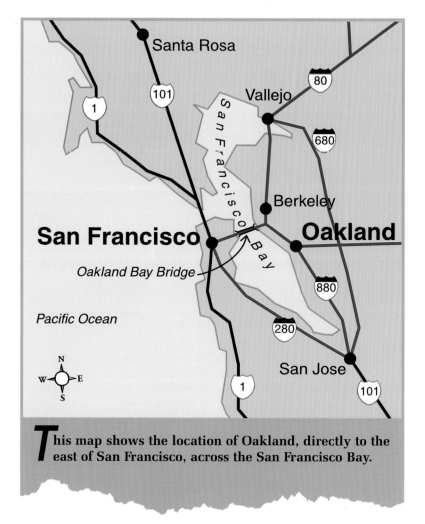

*T*his map shows the location of Oakland, directly to the east of San Francisco, across the San Francisco Bay.

called for more units to help fight the blaze. The air filled with smoke as wind whipped the flames and sent them in all directions. The hill below the brushfire erupted into a wall of fire. Wind picked up the burning embers and sent them flying into dry brush and trees. Those instantly burst into flames, too.

The San Francisco Bay Area had experienced five years of drought. Little rain had fallen, and vegetation was parched. In addition, temperatures in October had been unusually hot. They had reached ninety degrees Fahrenheit on the morning of the fire. The neighborhoods in the path of the flames were all located on heavily wooded hills full of dry brush and trees.

Large numbers of eucalyptus trees grew in many of the wooded areas. These trees were originally imported from Australia in the mid 1800s. They became very popular and were planted all over California. Because the trees contain a lot of oil, they are a serious fire hazard. The past years of drought made them even more dangerous than usual. In

*O*akland firefighter Lieutenant Phillip Bell makes sure that all the residents of an apartment complex have made it out safely.

Oakland neighborhoods, dry eucalyptus trees exploded as fire jumped from one to another.

Residents of the Oakland hills had been warned to cut down and remove dry vegetation, but many had not done so. In addition, most of the homes had wood-shingle roofs, which caught fire very easily. Burning embers were carried by the wind to the dry roofs. The fire quickly spread from one roof to another. Firefighter Joseph Cuff said, "When we first got there, trees were blowing up and baseball-size embers were blowing laterally with the wind—big chunks of them, and that's what caused the houses to catch fire."[2]

Fire trucks from all over the San Francisco Bay Area responded to the calls for help. The narrow, twisting roads leading into the Oakland hills were soon jammed with fire trucks trying to get up and cars trying to get down. Many of the roads were so steep and narrow that only one vehicle at a time could pass. Some of the fire trucks were too wide to move up the narrow roads.

The fire trucks that did get through were greeted with more problems. The Oakland fire department routinely used three-inch couplings to attach its hoses to the fire hydrants. Most other fire departments in the Bay Area used two-and-a-half-inch couplings, so they were unable to connect to the Oakland hydrants. In addition, the trucks that were able to connect found very low water pressure, because so much water was being used at the same time.

Many residents were trying to save their houses by

spraying water on the roofs to prevent them from catching fire. Also, in some burned houses pipes burst, sending water gushing out into the streets. To make matters even worse, many of the area water tanks were not able to pump water at all. The pumps ran on electricity, and power transformers were exploding as a result of the intense heat and flames. Without electricity, the pumps did not work.

The Oakland fire happened fast. The first alarm was called in at 10:58 A.M. By noon, seven hundred houses had burned, and the fire was just getting started.

Despite encountering many obstacles, the Oakland fire department worked tirelessly to fight the blazes erupting throughout the city.

CHAPTER 3

Escape!

On October 17, 1991, Raul and Dena Cruz and their five-year-old daughter, Conner, moved into a beautiful new house in the Oakland hills. Three days later, on Sunday, October 20, Dr. and Mrs. Cruz decided to go into San Francisco to buy some furniture. They left Conner with a baby-sitter. On their way into the city, Dena Cruz thought that the air seemed awfully hot and dry. "I told my husband it was fire weather. He thought I was crazy."[1] While they were driving, Dr. Cruz received an electronic page from home.

Using a cellular phone, the couple tried to call their baby-sitter, but the line was busy. After trying again and again with no success, they turned around and headed home. "We got within a few miles and saw a huge cloud of smoke and ashes falling," said Dena.[2] Where was Conner? Was she safe?

As they sped home, Dr. Cruz received a page from the hospital where he worked. After he phoned in, Dr. Cruz

This photo, taken from the Oakland hills, looks out across the bay to San Francisco in the distance. More than three thousand homes were destroyed by the fires on October 20, 1991.

was told that a neighbor had brought Conner to the hospital for safekeeping. When they picked up their daughter, Dena said to her, "I'll bet you were brave." Conner answered, "No, Mommy, I was scared."[3] The family's new house burned to the ground, but thankfully, everyone was together and safe.

On Sunday morning, at another house, Joy Harrison was in the backyard, fixing her five-year-old daughter's hair while her twins played with their baby brother. Inside, her husband, John Harrison, was relaxing and reading the newspaper. Suddenly the sounds of sirens filled the air. The family ran to the front yard to see what was going on. They saw thick, dark smoke in the distance, but thought it was just the brushfire from the day before that had started burning again.

When the smoke kept getting darker and closer, John Harrison climbed up on his roof and started spraying the shingles with water. While he was up there he could see homes across the canyon exploding into flames. A neighbor hollered to Harrison that they were surrounded by fire. Quickly, John got down and rounded up his four children. He put on their shoes and then backed the car out of the garage.

While John was putting the children in the car, Joy collected a few things from the house. She was going in for another load when fire roared over the hill above the Harrisons' house. One of the children screamed, "I don't want to die."[4] Joy finally stopped worrying about her possessions and ran to the car. As the Harrisons sped away,

John called his mother-in-law from the car and told her, "The house is gone."[5]

Smoke covered the sun and plunged the Oakland hills into near darkness. While residents were running away from the flames, firefighters were running toward them. Joseph Cuff from San Francisco Fire Engine Company No. 36 was at the leading edge of the firestorm. As his team was trying to put out a house fire, one of his men was blown off the roof by strong winds. Fortunately, the

Berkeley resident Jim Beatty stands hopelessly with a garden hose while homes in his neighborhood burn out of control.

*T*hough the Oakland fire department battled the flames fiercely, many houses were lost. Houses roofed with tile and other nonwood materials held up best.

man was not badly hurt, and the firefighters continued to battle the blaze.

Cuff said,

> We had lots of assistance from college-age kids who came up and asked if they could help . . . and they carried the stuff we needed up the hill for us. Some of the others—about ten or twelve of them—were moving hose under the direction of

Firefighter Zammarchi. They were doing all the leg work. If we had to run up and down that hill, we would have been exhausted and [of] no further use.[6]

After putting out the fire on the hill, Cuff's company headed for another blaze on the street above them. The students knew how to get up the hill, so they led the firefighters. "We couldn't have gotten up that hill without the

Many trucks from fire departments all over the Bay Area were used to put out hot spots throughout the city.

kids helping us. One of the barefoot kids took a small line and climbed up to the top of the hill and tied it around the tree," Cuff said.[7] The hill was so steep the firefighters had to grab fence posts to pull themselves up.

When they reached the top of the hill, there was chaos everywhere. Residents helped pull up the hoses while some of the students sprayed water down the hill. "Some other department had left five charged big lines [hoses

*W*orkers from Pacific Gas & Electric tried to repair downed power lines and damaged transformers. The lack of electricity to run water pumps seriously slowed firefighters' efforts to put out the blaze.

connected to hydrants] in the area, and the kids were attempting to put out the fire. There were about five kids on each line and they were shooting water down the hill, and that stopped the fire from coming up the canyon," related Lieutenant Dan Dente.[8]

After the fires on the hill were under control, the team began looking for other blazes to extinguish. It was hard to see just where fires were burning because of all the smoke. To help pinpoint the locations of the fires, two airplanes with remote sensing equipment joined in the fight. The NASA-Ames Research Center sent a C-130 transport plane and a high-altitude ER-2 plane aloft. The ER-2 sent back infrared images that helped ground crews locate the burning areas. The C-130 made twelve passes over the area and took videos of the scene below. After landing, the videos were taken by the California Highway Patrol to the fire command center.

Information from both planes allowed firefighters to see where the fires were hottest and in which direction they were moving. They also provided data for the firefighting aircraft that were trying to extinguish the flames from above. Ten fixed-wing tankers were dumping fire retardants onto the flames. They were joined in the air by twenty helicopters that dropped a foam mixture to help put out the fires.

Because the fire was burning in a relatively small area that was very hilly, only a few of the aircraft could fly at one time. According to the California Department of Forestry's Olis Kendrick, "We were close to maxed out on the number of aircraft we had there and could work effectively. As it was, only two tankers at a time could make runs, due to the constrained area and the need to intermingle with the helicopters."[9]

Kendrick said, "The heavy, dense smoke was also a problem. You can't fly through it. You have to be able to see going in and coming out, and you have to have an escape route. You have to plan around the smoke."[10] The wind was also creating hazardous flying conditions. The helicopters were better able to maneuver in the high

*B*ecause of the great number of helicopters in the air, only two airplane tankers at a time could be used to drop fire retardants and try to stop the spread of the firestorm.

winds, but their loads were not as large or effective as those dropped by the tankers.

Whereas the aircraft provided a great deal of help in fighting the fires, Kendrick emphasized that they could never have put out the flames alone. "We try to get in early and slow the advance until the ground forces get there." he said.[11]

CHAPTER 4

Triumph and Tragedy

While the battle was being fought from the air on that disastrous Sunday in October, firefighters on the ground continued their house-by-house assault. Fire units from all over the San Francisco Bay Area had responded to the crisis and were fighting side by side with the Oakland fire department.

Oakland battalion chief James Tracey established a command post in the fire zone.

> Before we started attacking the fire, I told company officers that I'd be watching the safety aspects of the operation. . . . I was prepared to have them drop everything and get out if that conflagration [fire] swept down on us, or if the wind changed direction.[1]

Tracey explained further: "I could hear PG&E [Pacific Gas and Electric] gas meters exploding in the area; they would just go 'pop' and then continue to burn with flames up to three feet high. I also heard tires exploding on automobiles parked inside burning garages."[2] Chief Tracey

*T*he smoke in the Oakland hills was so thick that it could
be seen from across the bay in San Francisco. The Bay
Bridge, pictured above, connects San Francisco with
Oakland and Berkeley in the east bay area.

watched a little smoke curling from the roof of a nearby house. He asked one of his crew members to hose down the roof. "I had him just hit it with as much as he could. It was like nothing. The roof was gone in five minutes, and the entire building gone in ten."[3]

The fire produced a huge cloud that extended above the Golden Gate Bridge and out over the Pacific Ocean. High winds carried burning embers as far as ten miles away and dropped them on the city of San Francisco. Ash falling from the black cloud piled up nearly an inch deep in some parts of the city. San Francisco residents began spraying their roofs with water to prevent them from catching on fire. All over the city, people gathered to watch the smoke and flames across the bay.

Firefighter King Strong described the scene when he arrived with Strike Team 1 to battle the Oakland blaze:

> The fire just kept on coming. . . . The fire jumped the street—it was on both sides of us—and it grew dark because the smoke was so intense it blocked out the sun. It felt like we were in the dead of night. . . . I saw embers flying—I actually saw little balls of fire flying all around in the air. You'd see fire flying everywhere—the intensity of it was incredible—I'd heard of firestorms, but this was the first time I was ever in one.[4]

At another location in the Oakland hills, the crew of Castro Valley Engine No. 5 was busy putting out a house fire when they heard a woman screaming for help on the other side of a fence. The fire crew knocked a hole in the fence and saw that the woman was cornered by flames.

Captain Brown said, "We knocked the fence down . . . and told her to retreat from the area. She was a photographer who was taking pictures of the oncoming fire when she got trapped."[5]

Firefighter Dwight Langford and his crew were trying to put out a small blaze when

> the wind picked up, and the fire came up the valley and completely surrounded us. . . . We were trapped, so we just wet each other down and lay down on the road to die. We held each other's hands and prayed. . . . We were definitely crying. It was a freak that the fire went right over us.[6]

Some were not so lucky. Sixteen-year Oakland police veteran John Grubensky saw five people running from a wall of smoke and flames. He yelled for them to get into his patrol car. When they were safely inside, he sped away from the fire, only to find the road blocked. Jumping out of the car, Grubensky shouted for everyone to get down and crawl. There was so much smoke that the only breathable air was close to the ground.

In spite of his efforts, Officer Grubensky and the five people he tried to save were killed. Their bodies were found together on a roadway. Grubensky's body had been burned so badly that he had to be identified by the serial numbers on his gun and badge. At least ten other people were also burned to death when their cars were overtaken by flames.

Oakland resident Bill Rogers was spraying his roof and watching the nightmare around him. Smoke alarms were screaming out their shrill warnings while cars and gas lines exploded. The fire had turned the sky an eerie red-orange color, and the heat was intense. When the house next door caught fire, Rogers tried to put out the blaze with his garden hose. He gave up the fight when the house exploded. He knew it was time to run.

Steve Hischler also knew it was time to run when the fire bore down on his house. "The trees were exploding. It looked like the Fourth of July. I think I lost everything; I think it's gone," he said.[7]

Baseball great Reggie Jackson, too, lost his house on

Many people died when their cars were suddenly engulfed in flames. The bodies of six people were recovered from this area.

that horrifying October day. When asked about the fire, he said, "It strips you. You're helpless. Defenseless."[8] When his house burned, Jackson lost many souvenirs from his years in baseball.

Joe Jorgenson almost lost his life while trying to save some valuables. He and his girlfriend were getting ready to go out when they looked out a back window and saw flames coming at them over the hill. "Five minutes later it was in the backyard," said

*B*aseball great Reggie Jackson was one of many who lost their homes in the Oakland fire.

Jorgenson.[9] They raced to get Jorgenson's mother, who was confined to a wheelchair. As they struggled to put Concetta Jorgenson into the car, the fire crept closer and closer. Jorgenson went inside to gather up some family heirlooms while his mother and girlfriend sped away in the car. He planned to escape on his motorcycle, which he also hoped to save from the fire.

With the flames bearing down on him, Jorgenson jumped on his Kawasaki and raced from the scene. Immediately in front of him was a wall of flames and burning embers. "I just prayed, closed my eyes and gunned it. . . . I could feel my shoulders and back on fire. There was a garden hose going on one of the lawns, so I

Many firefighters and police officers gave their lives battling the Oakland fire. Among those whose losses were mourned was Oakland fire department chief James Merle Riley, whose funeral is pictured.

stopped and doused myself."[10] Joe Jorgenson made it through, but he ended up in the hospital with second-degree burns over 20 percent of his body.

The Oakland firestorm lasted only one day, but that day seemed like an eternity to those who were there. By late afternoon, the wind died down and the fire was finally brought under control.

During the tragic day of October 20, 1991, more than three thousand homes were destroyed, five thousand people were left homeless, and twenty-six people lost their lives. Five of the dead were members of one family, who were trapped while trying to save some belongings. Total damages were estimated to be at $1.5 billion.

Alameda County coroner Joe Blackwell (left), coroner Grant Jenkins (center), and police officer Dan Mercado sift through ashes looking for bone remains of victims.

CHAPTER 5

Aftermath

When daylight dawned on Monday morning, a grim scene emerged. Now there were three square miles of ashes where houses had once stood on beautiful, wooded lots. Groves of trees had turned into acres of burned sticks. Abandoned cars lined the roads, their paint and tires burned off. Only chimneys stood as grim reminders that houses had once been there. Downed power lines and still-smoldering embers were everywhere. Firefighters worked for three more days, putting out all the hot spots before letting residents into the neighborhoods to see what was left of their homes.

Michael Rogers was escorted into the fire zone by two Red Cross workers. As a writer for *Newsweek* magazine, Rogers was used to reporting on disasters. This time was different, though, because he would be writing about his own destroyed house and neighborhood. As the trio climbed up a hill, one of the Red Cross workers whistled and said, "I've never seen anything like this."[1]

*A*mid the ashes and ruins, a single flag waves in the wind. Despite the destruction and trauma, people returned to their homes to rebuild and start over with their lives.

A*n Oakland police sergeant briefs local residents before they are finally allowed back into neighborhoods to inspect the remains of their homes.*

Rogers said,

> With the exception of two chimneys, the neighborhood was completely flat: no decks, no stairways, no second stories. It was still smoldering; the stench of smoke was everywhere. Down the street from me, police used German shepherds and bone sifters to search for physical remains. Other policemen stood guard against looters.[2]

When some people saw what was left of their houses, they broke down and cried. Neighbors hugged each other and vowed to rebuild. After shedding some tears, Dena Cruz said to her husband, Raul, "Okay, when do we start?"[3] They had lived in their house for only three days when it was destroyed by the fire. Dena Cruz was determined to rebuild on their wonderful lot that overlooked San Francisco Bay. Many other residents were also eager to get started on the cleanup and rebuilding. Within a few days, bulldozers moved in to knock down the charred remains, and dump trucks arrived to clear away some of the rubble.

Six months after the fire, Raul, Dena, and Conner Cruz moved into their new home. Their house was the first one finished on the street. The family had nothing—all of their belongings had been destroyed. Friends, neighbors, and even strangers pitched in and gave them dishes, silverware, and clothing. "Every day was like a birthday party," said Dena. "We got so much I had to give some away to the Red Cross."[4]

After getting settled, the Cruz family opened their doors to their neighbors, who were still in the process of rebuilding. People were free to stop in and have a cool drink or use the telephone. They all got better acquainted and helped each other through some very difficult times. Dena Cruz said that the fire changed their lives.

*M*any families' possessions were still smoldering when people came to investigate what was left of their homes.

Many fire victims found comfort in the company of family, friends, and neighbors.

"We learned what good friends we have and how kind even strangers can be. It's renewed my faith in people."[5]

In the months following the Oakland fire, many new foundations were poured, and builders hammered house frames into place. City officials made the process of getting building permits easier for victims of the fire. An empty grocery store in the area became the headquarters for the Community Restoration Development Center. Residents could get all of the permits they needed in one central location. The center helped speed up the process of restoring the Oakland hills to some kind of order.

Life gradually began to return to normal for the people who had experienced the devastating firestorm. Families moved into new houses and planted shrubs and trees to replace those that had burned. In spite of their joy at having a home again, some still thought about the family pets that had disappeared on that awful October day. About one thousand pets were still missing.

Kristine Barrett-Davis and Mark Davis wondered if their cat, Disney, had survived the fire. In an effort to try and find her, the couple sent pictures of Disney to the

*O*akland hills resident Barbara McFadden, standing in the rubble of her burned-out home, checks plans for a new home with architect Dave Boone.

area animal shelters. Kristine said, "I had dreams about her all of the time—that she had come back."[6]

Fourteen months after the fire, Kathryn Howell was visiting one of the shelters. She was a volunteer at the Firestorm Pet Hotline, an organization that tried to match up lost animals with their owners. One day she looked into a cage and saw a ". . . very friendly black-and-white cat who kept talking to me."[7] She took a picture of the cat and

*T*his thirty-five-foot cedar Christmas tree was planted by residents of the Oakland hills as a sign of hope for the rebirth of their neighborhood.

then returned to her office to compare it with photos in the file of missing pets. There were four possible matches. Howell called the four families and eventually determined that the cat might be the one lost by the Davises.

Kristine and Mark Davis raced to the center to see whether the black-and-white cat could possibly be Disney. Howell said, "When we took her out of the cage, she was shaking, but she was purring. I knew if it was her there would be a dark birthmark, sort of a zigzag, inside her mouth—and there it was!"[8] After surviving on her own for a year, the cat would finally have a home again. Holding Disney tightly in her arms and gently stroking her, Kristine said, "It was like a miracle that she came out of this trauma."[9]

Today, many of the homes that were destroyed in the Oakland fires have been rebuilt. Tile roofs are visible where once there were flammable, wood-shingle roofs. Dry brush and trees have been removed to help prevent future fires. Residents have planted trees and shrubs to replace those lost in the fires, but now trees must be planted at least fifty feet away from the houses.

Roads into the hills have been widened so that more than one vehicle at a time can pass. Now large fire trucks can easily drive in and out of the neighborhood streets. High up in the hills, where no roads go, there are still areas of burned tree trunks visible from below. The barren, blackened sticks stand as a grim reminder of the death and destruction that visited the Oakland hills on October 20, 1991.

Other Fires in the United States

DATE	PLACE	OUTCOME
October 8, 1871	Peshtigo, WI	More than 1,100 dead and 2 billion trees burned in forest fire.
October 8, 1871	Chicago, IL	250 dead; 17,450 buildings burned.
December 5, 1876	Brooklyn, NY	More than 300 dead in theater fire.
September 1, 1894	Minnesota	480 dead and six towns destroyed in forest fire.
December 30, 1903	Chicago, IL	602 dead at Iroquois Theatre.
June 15, 1904	New York, NY	1,030 dead in excursion boat fire on the East River.
March 4, 1908	Collingwood, OH	175 children dead in schoolhouse fire.
March 25, 1911	New York, NY	145 women dead at Triangle Shirtwaist Factory; investigation improved fire codes in all city buildings.
April 21, 1930	Columbus, OH	320 inmates dead at Ohio State Penitentiary.
March 18, 1937	New London, TX	294 dead in schoolhouse explosion and fire.
November 28, 1942	Boston, MA	491 dead at Coconut Grove nightclub.
July 6, 1944	Hartford, CT	168 dead in fire and ensuing stampede in the main tent of Ringling Brothers Circus.
December 7, 1946	Atlanta, GA	119 dead at Winecoff Hotel.
April 16, 1947	Texas City, TX	516 dead from explosion and fire.
December 1, 1958	Chicago, IL	95 dead at private school.
June 30, 1974	Port Chester, NY	24 dead in discotheque fire.
May 28, 1977	Southgate, KY	167 dead at Beverly Hills Supper Club.
November 21, 1980	Las Vegas, NV	84 dead at MGM Grand Hotel.
September 4, 1982	Los Angeles, CA	24 dead in apartment fire.
March 25, 1990	New York, NY	87 dead at Happy Land Social Club.

canyon—A deep, narrow valley between high steep cliffs or mountains.

conflagration—A large fire.

coupling—A mechanical device that connects two parts.

ember—The red-hot remains of a fire.

eucalyptus—A tall evergreen tree valued for its timber, gum, and oil.

extinguish—To put out or smother a fire.

fire retardant—A chemical used to slow the spread of a fire.

firestorm—An intense fire spread over a large area by strong winds.

hydrant—A large pipe with a valve for drawing water from a water main.

infrared—Rays of light that are beyond the spectrum visible to the human eye.

parched—Dry; lacking much-needed moisture.

smolder—To burn slowly without flames.

tanker—An airplane used for transporting and dropping fire retardant chemicals.

transformer—A device used to change electricity between circuits.

vegetation—Plant life.

Chapter 1. Sunday, October 20, 1991

1. June Jordan, "The Fire This Time," *The Progressive*, January 1992, p. 11.

2. Ibid.

3. Ron Arias and Tony Chiu, "California Burning," *People Weekly*, November 4, 1991, p. 45.

Chapter 2. The Beginning

1. Captain Donald R. Parker, *The Oakland-Berkeley Hills Fire: An Overview*, p. 1, n.d., <http://www.sfmuseum.org/oakfire/overview.html> (January 22, 1998).

2. *Strike Team Operations: Reata Place and Chabot Road*, p. 2, n.d., <http://www.sfmuseum.org/oakfire/e3and36.html > (January 22, 1998).

Chapter 3. Escape!

1. Charles Salsberg, "Homecoming," *Redbook*, October 10, 1992, p. 208.

2. Ibid.

3. Ibid.

4. Paul Witteman, "How Do You Rebuild a Dream?" *Time*, November 4, 1991, p. 24.

5. Ibid.

6. *Strike Team Operations: Reata Place and Chabot Road*, p. 2., n.d., <http://www.sfmuseum.org/oakfire/e3and36.html> (January 22, 1998).

7. Ibid, p. 3.

8. Ibid.

9. Richard O'Lone and Bill Henderson, "Airborne Imaging, GPS Aid Aircraft, Firefighters in Battling California Blaze," *Aviation Week and Space Technology*, October 28, 1991, p. 71.

10. Ibid.

11. Ibid.

Chapter 4. Triumph and Tragedy

1. *Broadway and Ocean View Operations: Oakland Division G*, p. 3., n.d., <http://www.sfmuseum.org/oakfire/broadwayops.html> (March 18, 1998).

2. Ibid.

3. Ibid, p. 4.

4. *Strike Team 1 Operations: Eustice and Golden Gate*, p. 1, n.d., <http://www.sfmuseum.org/oakfire/stl.html> (March 19, 1998).

5. *Alvarado Road Fire Operations*, p. 4, n.d., <http://www.sfmuseum.org/oakfire/alvarado_pt.1.html> (December 26, 1998).

6. Ron Arias and Tony Chiu, "California Burning," *People Weekly*, November 4, 1991, p. 47.

7. *The Oakland Firestorm: A History*, p. 1, n.d., <http://www5.ced.berkeley...aegis//eastbay/shist.htm> (December 12, 1997).

8. Arias, p. 48.

9. Ibid, p. 47.

10. Ibid.

Chapter 5. Aftermath

1. Michael Rogers, "No More Home Sweet Home," *Newsweek*, November 4, 1991, p. 34.

2. Ibid.

3. Charles Salzberg, "Homecoming," *Redbook*, October 1992, p. 208.

4. Ibid.

5. Ibid.

6. "The Cat Came Back," *People Weekly*, February 8, 1993, p. 48.

7. Ibid.

8. Ibid.

9. Ibid.

Arias, Ron. "California Burning." *People Weekly*, vol. 36, November 4, 1991, pp. 44–49.

"The Cat Came Back." *People Weekly*, vol. 39, February 8, 1993, p. 48.

"Jackson, Henderson Among the Victims of Oakland Fire." *Jet*, vol. 81, November 11, 1991, pp. 7–8.

Knapp, Brian. *Fire*. Austin, Tex.: Raintree Steck-Vaughn Publishers, 1990.

Lambert, David. *Fires and Floods*. New York: New Discovery Books, 1992.

Ritchie, Nigel. *Fire*. Brookfield, Conn.: Millbrook Press, Inc., 1998.

Robbins, Ken. *Fire*. New York: Henry Holt, 1996.

Rogers, Michael. "No More Home Sweet Home." *Newsweek*, vol. 118, November 4, 1991, pp. 34–35.

Salzberg, Charles. "Homecoming." *Redbook*, vol. 179, October 1992, p. 208.

Sullivan, Margaret. *Firestorm!: The Story of the Nineteen Ninety-One East Bay Fire in Berkeley*. Berkeley, Calif.: City of Berkeley, 1993.

Wood, Leigh. *Fires*. New York: Twenty-First Century Books, Inc., 1995.

Ashland Fire & Rescue: Remembering the Oakland Conflagration
<http://ashland.or.us/Page.asp?NavID=64>

FEMA—Project Impact Volunteers Retrofitting 40 Homes in Oakland, California
<http://www.fema.gov/regions/ix/1998/r9_027.shtm>

Map of Oakland, California, Fire Stations
<http://www.qsl.net/orca/ares/OAKFIRE.HTM>

Oakland Fire Department
<http://www.oaklandnet.com/oakweb/fire/index.html>

Oakland Fire Weather
<http://oaklandnet.com/oakweb/revised/FWX.htm>

Oakland Hills Fire Storm: Remote Sensing and Emergency Management
<http://geo.arc.nasa.gov/sge/jskiles/fliers/all_flier_prose/
oaklandfires_brass/oaklandfires_brass.html>

San Francisco Fire Department Response to the Oakland-Berkeley Hills Conflagration, October 1991
<http://www.sfmuseum.org/oakfire>